LITHIUM MINING IN KACHI CATAMARCA ARGENTINA

World-Class Kachi Lithium Project with Lake Resources

JUDITH WALTER

All right reserved. No part of this publication may be reproduced, distributed, transmitted in any form or by any means, including photocopy, recording, or other electronic or mechanical methods without the prior written permission of the publisher except in the case of a brief quotation embodied in critical reviews and certain other noncommercial uses permitted by copyright law.

Copyright ©2022 JUDITH WALTER

Table of Contents

INTRODUCTION

CHAPTER ONE
- Catamarca Tres Quebradas Lithium Project

CHAPTER TWO
- New Lithium Mining Technology Could Give Argentina a Sustainable Gold Rush

CHAPTER THREE
- Lilac Solutions Continues to Advance World-Class Kachi Lithium Project with Lake Resources

CHAPTER FOUR
- Argentine environmentalists are interested

CHAPTER FIVE

- Argentina is currently at the forefront of a worldwide race for lithium

CHAPTER SIX

- Brines recovered from the salt lake in Kachi are kept in a well.

CONCLUSIONS

INTRODUCTION

New Lithium Mining Technology Could Give Argentina a Sustainable Gold Rush Argentina -lithium-mining-1 Lake Resources breaks the ground for its Kachi lithium mine in Catamarca,

The Vasquez brothers aren't used to visitors. Their property is located in the Puna, a wide plateau region in the Andes Mountains, approximately 12,500 ft above sea level, and a full day's trip to the nearest city.

The landscape, in the Argentine province of Catamarca, is severe and completely uninhabited; fluffy, big-eyed llamas traverse a miles-wide plain between mountains. Only a few shrubs scatter the ground, glittering yellow-green Technicolor under the close light.

But one day in 2016, a tall man in his 50s, speaking badly Australian-accented Spanish, strolled up to the Vasquezes' remote property. He told them that a few kilometers away, under the otherworldly surface of the plateau, lurked enormous amounts of lithium—the white metal necessary to create the batteries needed to power electric automobiles and other clean energy technology—and he had the plan to obtain it.

Such immigration arrivals are typically bad news in the Andean highlands, which span across areas of Argentina, Chile, Bolivia, and Peru. Over the past 30 years, North American, European, and Asian mining enterprises have crowded the region to dig up its enormous quantities of copper, zinc, silver, and lithium, of which 59 percent of the world's known reserves are here. But mining interferes with one of the world's driest ecosystems: sections of the Puna may go years without rain, and populations here rely on a rare network of rivers and salt lakes, fed by subterranean water storage built up over thousands of years. Since the 1990s, mining businesses in northern Chile have drawn water out of salt lakes to collect the lithium lurking underground. The impact on aquifers is still poorly understood, but courts and communities in Chile believe mining has lowered groundwater levels, affecting the existence of entire Indigenous populations.

Argentina is today at the forefront of a global hunt for lithium. The country, which alone accounts for 21 percent of the world's reserves, has just two mines in operation today, but 13 more are planned and dozens more are under consideration—the world's greatest lithium project pipeline. Argentina's embryonic boom reflects a remarkable transformation in the lithium market: a few years ago, lithium was a very specialized commodity, used to manufacture glass, ceramics, and lubricants. Most of it was taken from well-established mines in Chile, Australia, and China. But with the global energy transition anticipated to trigger a 40-fold spike in lithium demand between now and 2040, according to the International Energy Agency, mining firms are vying to guarantee supplies in less developed nations, such as Argentina and

Bolivia, as well as Serbia and Mali. Many worried local populations in these places, notably the Argentina Puna, home to some 50,000 people, are destined to face the same destiny as those in Chile: having their resources pillaged and lands devastated to service the markets of wealthier countries—neocolonialism dressed up as a green revolution.

But Stephen Promnitz, an Australian mining executive, assured the Vasquezes that he had the tools to gather lithium while safeguarding their birthplace. "He was pretty polite," recalls Florentín Vasquez, a nice 38-year-old with a floppy black sun hat, standing with his two brothers beside a half-built adobe house in March 2022. "He says that they're not going to require as much water as prior projects—that they have a revolutionary technique that has never been tried before in Argentina."

CHAPTER ONE

Catamarca Tres Quebradas Lithium Project.
known as the 3Q lithium project, is a high-quality lithium brine project being developed by Neo Lithium in the Catamarca Province of Argentina. The anticipated capital investment for the project is $319m.

The project is predicted to generate an average of 20,000 tonnes (t) of battery-grade lithium carbonate-equivalent (LCE) a year throughout its planned mine life of 35 years.
The mining and environmental authorities of Catamarca authorized the 3Q project in October 2016, while the pre-feasibility study was completed in March 2019. A feasibility analysis of the project is planned to be completed in the first half of 2020.

Neo Lithium is presently working on the pilot plant, which is in the final phases of commissioning.

Tres Quebradas lithium project location, geology, and mineralization

Located in the southwestern region of the Catamarca Province of Argentina, the project sits roughly 25 km from the border with Chile. Extending over 35,000ha, the project contains a solar and a 160km² brine reservoir complex.

The unique high-grade lithium brine resource is found in the Lithium Triangle on the Puna Plateau near the main Andean Cordillera. The Lithium Triangle is home to many high-altitude salt lakes and salt flats with increased lithium contents.

Tres Quebradas lithium project reserves

The proved and probable reserves of the 3Q project were assessed to be 1.3Mt of LCE grading 794mg lithium, as of March 2019.

Mining at Tres Quebradas

Conventional evaporation pond operation is envisaged to remove the brine from the project. Brine from the evaporation ponds will be transferred to producing wells by self-priming pumps.

Major manufacturing equipment needed for the project includes trucks, a hydro crane, two high-density polyethylene (HDP) welders, and submersible electric pumps. The servicing equipment will comprise bulldozers, frontend loaders, and dump trucks.

Brine processing for Tres Quebradas lithium project Standard, low-cost solar evaporation technology will be applied to generate lithium carbonate from the Tres Quebradas project.

Brine from the producing wells will be piped to the solar evaporation pre-concentration ponds. NaCl (halite) and KCl (sylvite) will be removed from the brine after an approximately 120-day retention period.

The concentrated brine will then be delivered to the calcium chloride ($CaCl_2$) ponds and thickeners to remove the bulk of the calcium precipitating antarcticite. The brine with a lithium content of 3.5 percent will be transferred to the brine processing facility.

Brine processing will be undertaken in four phases at two sites - Fiambalá (100 km from Salar 3 Quebradas) and Recreo (465 km from Fiambalá).

At the Fiambalá facility, the solvent will be extracted from the brine to remove residual boron. It will be followed by a sulfation procedure to remove leftover calcium by adding a saturated solution of sodium sulfate.

The brine will be combined with mother liquor and mild soda ash at the Recreo plant to eliminate residues of calcium and magnesium. Soda ash and heat will be used to precipitate lithium carbonate, before continuing to drying and packing procedures.

The sulfation plant is projected to recover 92 percent of LCE, while the carbonation plant is predicted to recover 85 percent.

Infrastructure facilities

The lithium project may be accessible via an access road from the Ruta Nacional 60 Highway, which sits within 60km of the project.

The power supply for the 3Q project will be supplied by a hybrid system incorporating a photovoltaic (PV) and diesel production system. The PV and diesel plants are anticipated to supply 2.6MW and 8.8MW, respectively.

Fresh water is intended to be provided from a shallow well situated roughly 4km from the mining site.

Contractors involved in Groundwater Insight provided the NI 43-101 technical report and the mineral resource estimate of the project. Groundwater Insight was helped by Hidroar and Martin and Miguens with the project studies.

Groundwater Insight was also recruited jointly with GHD Chile (GHD) to write a technical report detailing the PFS. Other participants in the technical report development were ILLA Research Group, Golder, and G&T Ingenieria.

GHD was responsible for the compilation of the preliminary economic assessment (PEA) report of the Tres Quebradas lithium project

CHAPTER TWO

New Lithium Mining Technology Could Give Argentina a Sustainable Gold Rush.

Lake's flagship Kachi Project is situated in Argentina's Catamarca Territory at the southern end of the Lithium Triangle, a world-renowned province responsible for 40 percent of worldwide lithium output.
Lake's flagship Kachi Lithium Project.
Using direct extraction technology offered by Lake's technology partner, Lake hopes to develop at Kachi a sustainable, high purity lithium carbonate product capable of fetching premium price from battery cathode producers.

Importantly, the direct extraction technique is an ecologically benign technology as practically all the water (brine) is returned to its source, enabling the EV and battery sector to be a sustainable and continuously high-quality source of lithium for their supply chains.

Kachi encompasses 705 sq km of leases and leases applications across a brine-bearing basin 20 km long, 15 km wide, and 400-800m deep.

In November 2018, Lake reported a first JORC resource at Kachi of 4.4 million tonnes (Mt) of contained lithium carbonate equivalent (LCE), and an exploration target ranging between 8-17 Mt of LCE. This ranks Kachi as one of the world's top 10 brine resources.

In April 2020, Lake revealed convincing pre-feasibility study (PFS) findings for Kachi, revealing its potential to become a long-life, low-cost operation with an annual output goal of 25,500 tonnes of battery-quality lithium carbonate utilizing direct extraction technology. The findings demonstrated a high-margin project, with an EBITDA margin of 62 percent, combined with competitive capital and operational expenses.

The next steps for Kachi include delivering product samples from Lilac's pilot plant to potential off-takers; targeting lower upfront and operational costs, such as through using environmentally friendly solar power; and further resource development to extend the project's life, in addition to financing and off-take discussions.

Disruptive" technology
Currently, lithium extraction from salt brine is a time and labor-intensive process that creates enormous quantities of trash.
Brine from subsurface deposits, mostly located in the South American region known as the "Lithium Triangle" (Chile-Bolivia-Argentina), is pumped to the surface and processed through a succession of vast evaporation ponds with chemical treatments applied.

Lilac's ion-exchange method will enable brine to be returned underground, decreasing the overall environmental effect.

"Lilac's technology has taken a non-mining tech solution which saves operational costs and enhances lithium recovery from our brines," Lake Resources managing director, Steve Promnitz said in the release.

Lithium prices increased throughout the last year, lately hitting their highest levels since mid-2018. Battery-grade lithium carbonate prices have climbed more than 20 percent in the first half of September and are currently up 188.9 percent and 215 percent, respectively on the Chinese domestic market this year, data from Benchmark Mineral Intelligence indicates.

Demand for lithium is predicted to surge 26.1 percent or around 100,000 tonnes of lithium carbonate equivalent to a total of 450,000 tonnes, turning the market into a deficit of 10,000 tonnes.

Lilac Lithium Extraction Technology

As well as funding and offtake assistance, Lake has proved its technology partner Lilac Solutions' direct lithium extraction technique is scalable and economically efficient.

Unlike typical lithium brine production, Lake aims to adopt Lilac's direct lithium extraction technology, which does not involve evaporation, uses less water, and has a reduced environmental effect.
Lilac is gaining a 25 percent stake in Kachi and is giving its technology, engineering staff, and an on-site demonstration facility.
The direct lithium extraction technology is cheaper and gives better lithium recovery rates than older approaches.

"The process is modular generating high purity lithium and can be scaled up fast from pilot to commercial phases — this ownership participation assures a speedy development of the Lilac technology at the Kachi site," Mr. Promnitz added.

"With the Lilac technology, we can effectively produce the enormous amounts of high-quality lithium compounds required by battery makers."

"Importantly, this will be done in a manner that is environmentally friendly," he continued.

CHAPTER THREE

Lilac Solutions Continues to Advance World-Class Kachi Lithium Project with Lake Resources

OAKLAND, Calif., July 6, 2022 — Lilac Solutions, a lithium extraction technology business, reiterates its commitment to swiftly developing the Kachi lithium project in conjunction with Lake Resources. Lilac has designed and constructed a lithium extraction process facility for the Kachi project and successfully delivered this equipment to Salta, Argentina in April. This factory is intended to create a lithium chloride concentrate, with a production capacity of 40 tonnes per year of lithium carbonate equivalents. Lilac aims to put this plant on-site in July provided Lake Resources completes the development of the on-site infrastructure, and certain results from on-site operations are thus anticipated by the end of 2022.

Lilac has invented a novel ion exchange technique to boost the production of lithium from brine deposits. Electric car companies are experiencing high pricing and shortages for lithium raw materials as traditional methods for lithium manufacturing struggle to keep pace with demand. Based on extensive test work and FEL-2 engineering studies done with Hatch Engineering, Lilac thinks its technology is well-suited to the Kachi project brine resource and is convinced that the on-site process facility will be successful. Lilac is strongly committed to working with Lake Resources to develop the Kachi project into commercial production as one of the world's biggest, most modern, and most sustainable lithium projects.

"Lilac's custom-built lithium extraction process plant for the Kachi project arrived in Argentina in April, and the Lilac team is fully prepared to install and commission it on-site" stated Dave Snydacker, CEO of

Lilac Solutions. "We are convinced that the on-site piloting at Kachi will mirror earlier test work and justify our view of Kachi as one of the biggest and best-positioned lithium projects worldwide. We feel that Lake's planned restructuring of its senior leadership, including bringing in management with greater expertise in late-stage project development, is in the best interest of the project."

Lake Resources continues to pursue the Kachi definitive feasibility study ("DFS") with Hatch Engineering, and Lilac has supplied replies to all information requests made by Hatch. Lake and Lilac aim to create a DFS committee to finish this analysis in tandem with the rearrangement of Lake's management team.

David Alt, Lilac's Vice President of Engineering, Lithium, who spent more than 25 years with Fluor Corporation, spoke on the development of the Kachi project and the DFS. "The Lilac team's rigorous test-work on Kachi brine offers a robust basis for the DFS. Lilac has helped both the Lake Resources and Hatch teams to integrate outcomes of this test-work into the overall flowsheet and DFS research, and we look forward to completing a successful DFS later this year and moving into project execution," stated About Lilac Solutions

Lilac Solutions is a lithium extraction technology firm situated in Oakland, California. Lilac has created a proprietary ion exchange technique that permits the production of lithium from brine resources with great efficiency, minimum cost, and ultra-low environmental imprint. Lilac's objective is to increase worldwide lithium production to help the electric car sector and energy revolution.

A few kilometers away from the brothers' property, Lake Resources, the mining firm Promnitz started six years ago, is currently laying the framework for a salt lake lithium mine—dubbed Kachi. Using technology from California-based start-up Lilac Solutions, the business plans to start manufacturing lithium carbonate (the metal compound that battery makers purchase) in 2024, delivering 50,000 metric tons a year by 2025.

Traditional lithium mines rely on a simple two-year-long evaporation process to separate lithium from the salty brine, allowing massive amounts of water to escape; by contrast, in a few hours, Lilac's technology can recover up to twice as much lithium and return "virtually all" of the salt water to its aquifer, according to Promnitz.

Flamingos were discovered in the saline lagoon near where Lake Resources' Kachi project will be constructed. Mining growth in the Andes might impact a fragile ecology This sort of method is known as immediate lithium extraction or DLE, and Kachi is one of the world's most sophisticated projects to employ it. The technique is untested, though. Even with major lithium producers researching DLE, including the world's biggest producer, North Carolina's Albemarle Corp., experts say it has failed to get from the lab to the field. Some investment organizations have raised severe reservations that Lake Resources will make it work at scale on its ambitious timeframe. The corporation has also endured internal upheaval: in June, Promnitz unexpectedly announced his departure. He tells TIME changes to the company's leadership were "anticipated" ahead of commencing construction at Kachi, while Lake's chair Stu Crow claims the resignation "was for strictly personal reasons." Despite the turbulence, Kachi remains a vital test case for DLE.

The economic incentives to roll out the technology couldn't be clearer. At the present, you can't power electric cars or store renewable energy without lithium. A frenzy to secure supply has sparked a roughly 500 percent surge in the price of lithium carbonate in the past 12 months—though experts think the pricing constraint may ease owing to a flood of new lithium investments.

Proponents argue DALE's speedier, more efficient technique is important to ramp up lithium production and avert fatal bottlenecks in the energy shift, hampering the battle against climate change. U.S. Energy Secretary Jennifer Granholm has termed it "a game changer" for the battery sector.

David Snyder, Lilac's CEO, argues it must succeed: "Conventional players have not been capable of supplying fresh supplies and the levels necessary for electric vehicles," he adds. "So by 2030, either there's a disaster in the electric car market, or the lithium sector has been fundamentally transformed."

Bolivia's Marxist government, which possesses the world's biggest lithium resources but has dreaded the social and environmental effects of using them, seems to view DLE as the answer. In June, following a trial program investigating lithium recovery rates and water consumption, authorities declared that six DLE businesses, including Lilac, would be permitted to compete for lithium contracts there.

But amid the rush for lithium, Argentine environmentalists argue that a fairer future is far from assured. Most mining businesses aren't waiting for the adoption of cleaner tech. And projects like Kachi, which promise "cleaner lithium," have yet to show they can run without sapping freshwater supplies or affecting a little-understood environment. The uncertainty hangs on the Vasquez brothers, who have lived on this property for three generations, and expect to pass it on to their children one day. "People from outside may come and tell you, 'Don't worry nothing will happen,'" Florentín explains. "But we're the ones at risk."

When construction is complete in two years, there will be additional extraction wells and covered tanks for the ion-exchange process vital to DLE's efficiency. "We'll put the brine in those tanks for only three hours," Promnitz says, squinting at the sun behind his glasses. Inside, lithium atoms will break from the water molecules, and bond instead with tiny ion-exchange beads produced by Lilac Solutions.

The beads are then fished out from the brine and washed with a strong acid to separate lithium chloride. Meanwhile the brine—around 800 metric tons of it per metric ton of lithium carbonate produced—can be returned to the aquifer, Promnitz says. In theory, that should prevent groundwater depletion as reported by communities in Chile. Some freshwater, however, is later used in the process of turning lithium chloride into lithium carbonate, to be shipped to companies that make battery cathodes.

CHAPTER FOUR

Argentine environmentalists are interested, but not reassured, by Lake and Lilac's promises to cut water usage. "They are guaranteeing us it would have a reduced impact," says Patricia Marconi, a Catamarca-based researcher with the YU CHAN Foundation, a regional conservation organization. "But they don't have anything published." She alleges Lake has refused to share information with her and her coworkers. The corporation has not yet produced its environmental impact assessment study, generating doubt. (Lake Resources chair Stu Crowe tells TIME the study is still being completed and is slated to be released in the third quarter of 2022.)

Brines recovered from the salt lake in Kachi are kept in a well. Sebastián López Brach for TIME

Eva Vasquez was born in the region where the Kachi project is being created. Sebastián López Brach for TIME

Marconi's worries revolve around two issues: First, how will reinjecting vast volumes of brine into the aquifer alter the geological formations below the surface? In a 2018 publication, a group of Argentina-based academics stated that reinjecting the previously lithium-rich brines into the lakes is "a very risky oversimplification" of the possible environmental consequences of the practice. According to the report, treating the brines to remove lithium might change their pH and introduce trace quantities of foreign chemicals. And firms may have to inject leftover brine at other sites than where they extracted it, simply to prevent diluting the lithium concentration where they're extracting.

The second issue is, how much fresh water will be used in the latter phases of the process? Marconi cautions that this water will likely be so-called fossil water—drawn from aquifers locked below for thousands of years that are not supplied quickly enough by today's rains to be restored.

All of this, Marconi argues, will have an unknown influence on the little-studied geology and sensitive biology of the salt lakes. In their zeal for lithium, she argues Argentina's national and provincial governments are neglecting to complete the research necessary to predict the implications of huge mining development, turning the country into a "free for all" for the lithium business. "If we were truly taking seriously the principle of not engaging in environmental systems that we don't understand, there wouldn't be 20 corporations investigating in Catamarca. There would be 20 study teams examining what's going to happen," she adds. "Because the damages are irreversible."

Catamarca has caused to question DLE's environmental claim. Three hours north of Kachi—a teeth-chattering journey over rough mountain roads—the yellows and greens of the Puna's summer flora are abruptly broken by a startling expanse of black ground. This is the valley of the Trapiche River, a water supply for the huge Hombre Muerto salt lake. In 1997, Livent, a Philadelphia-based lithium mining company—a critical supplier for both Tesla and BMW—built a small dam at the point where the river flows into the salt lake. The dam concentrates the freshwater for use in Livent's mine, which now can generate up to 20,000 metric tons of lithium per year. On TIME's visit in March—the final month of the Puna's rainy season—a trickle of water a few feet wide went beyond the dam, across a parched, scorched grassland.

The Trapiche River's course is blocked by a dam erected by Livent near the Hombre Muerto salt lake. Sebastián López Brach for TIME

The operation is Argentina's oldest lithium mine—and it's also the only one in the western hemisphere to deploy a variant of

DLE at scale. Its procedure is a hybrid: lithium brines are allowed to evaporate in pools but for "significantly less time," according to Livent than in standard methods, decreasing saltwater loss. The brine is then processed through a DLE process and subsequently, Livent adds, "most" of the salt water is returned to "the neighboring Salar habitat." Later, fresh water from the Trapiche River is used to separate the lithium. The firm did not release any data on brine consumption, but asserts it has "not contributed to a drop in brine or water [in the two decades it has worked] at the Salar."

Román Guitián blames Livent for the valley's ruin. Guitián grew up adjacent to the river, in a tiny Indigenous hamlet made up of his family and a few others. Before the mining began when he was 17, they used to get salt from Hombre Muerto, and farmed llamas, goats, and sheep on the valley's vegetation, says Guitián, standing at the edge of the salt lake next to a beat-up 4×4 that he employs to take visitors around the highlands. "It was lovely. But now there are no animals since it's completely dry."

Signs strewn along the river proclaim an initiative to rehabilitate the valley via reforestation and new irrigation systems, which Livent initiated last year with a regional NGO, the Eco-Conciencia Foundation. And yet, in early March, during a ceremony at Hombre Muerto attended by Argentine politicians and Tesla executives, Livent announced intentions to increase the plant's lithium production capacity by the end of 2023. The business is also planning two more extensions, aiming to raise its total capacity to 100,000 metric tons by 2030. In investment documents, Livent promises that the company would combine "reuse" and "recycling" to decrease its fresh water usage in the future. But it also adds that the latter stages of growth will utilize "a more typical pond evaporation-based process."

Indigenous activist Román Guitián poses beside the Hombre Muerto salt lake. Sebastián López Brach for TIME\Catamarca governor Raúl Jalil says the region has learned from the repercussions on the Trapiche River. "There are things which maybe went the wrong way in the past, but we are fixing those," he adds. "We are taking greater control now." Companies are directly forced to undertake monthly environmental monitoring, and if concerns crop up, projects will be stopped, Jalil said. The Levant claims it has already constructed monitoring stations at both the Trapiche and Los Patos rivers "to check water levels, recharge rate, and water chemistry to enable us to utilize water sustainably."

Jalil, though, says he does not seek to limit the number of lithium projects permitted in the province, or block conventional, water-intensive technologies in future mines, unlike Bolivia's government. "All enterprises, from agriculture to tourism, have an environmental effect, and we can't overhaul the entire global energy system without mining," he argues. "The path ahead is to lessen the effect more and more, via technology, innovation—at the same time as [continued] extraction." He wants Kachi to be "a leading case" there.

Promnitz thinks Kachi will utilize water more effectively than Livent has: Livent's worldwide freshwater consumption for 2020—based mostly on the Hombre Muerto project, its sole operating lithium carbonate mine—was 72.9 metric tons for every metric ton of lithium carbonate equivalent (LCE) produced. In the lab, Lilac claims it consumed 18 metric tons of water for every metric ton of LCE. Lake believes the rate at Kachi will be much below that, and that more water savings would come from plans to utilize brackish and recycled water.

Promnitz, who started his career as a geologist for mining industries, also thinks the brine reinjection procedure is not as hazardous as environmentalists worry. The process is similar to one that has been used for decades to enhance oil recovery in the U.S. shale sector, he says, except that the looser sediments of the salt lakes, compared to the tighter formations under oil fields in Texas and Louisiana, should make it less important where the brine is reinjected. "It's not like [the brine] lives in just one specific structure. We're merely pulling the lithium out, and putting back precisely the same brine that was there before."

In discussion, Jalil and other local leaders in Catamarca seem considerably more concerned with mining's economic possibilities than its environmental hazards. Argentina's Northwest has always been a backwater, getting little investment from either Buenos Aires or elsewhere. Politicians view mining as an opportunity to alter that:

corporations such as Livent have paid for roads and bridges in areas that were until recently inaccessible during poor weather; Lake is in negotiations to recruit local people and contract services, such as laundries, with local businesses. Though leery of unmet promises, the mayors of both El Peñón and Antofagasta de la Sierra, Kachi's two nearest towns, express variants of the same phrase: "If the firm expands, the municipality should too."

It's reasonable that local authorities would adopt that approach, says Juan Carrizo, head of the Eco-Conciencia Foundation, which tries to "resolve socio-environmental conflicts" surrounding mining. "It's simple to protect the environment from areas like Buenos Aires, where you have the internet, roads, gas, and transport," he argues. "But here, the growth of the community is also on the table."

A discussion over mining plays out on the streets of Antofagasta de la Sierra. Proudly displayed bumper stickers say "I'm a huge fan of mining," while graffiti begs "hands off our land and water." Florentín, whose property stands near the Kachi site, feels conflicted. He argues that if environmental concerns did develop from the project, he wouldn't be sure what to do. "We can attempt to push the corporation to leave ... but I don't believe we want to do that," he replies softly. "There are a lot of individuals around here that need a job, and they come and tell me that. So I feel sort of cornered."

We are experiencing a worldwide mineral mining boom. Carbon-reducing technology like wind turbines, solar panels, and electric automobiles demand a bigger volume and a more diversified variety of minerals than their dirtier equivalents. Already, since 2010, the International Energy Agency stated, "the average quantity of minerals required for a new unit of power production capacity has climbed by 50 percent as the percentage of renewables has risen." That provides the global mining sector a shot at a rebrand: from environmental villain to climate savior.

But in an age of heightened environmental consciousness, people are also fighting back against mining operations. In January, following weeks of huge demonstrations, the government in Serbia—thought to be home to Europe's greatest lithium deposits—shelved a $2.4 billion project sponsored by mining giant Rio Tinto, which could have given 90 percent of Europe's current lithium demands. Though Rio Tinto is now seeking to continue discussions with the government, which won re-election in April, the experience is a terrible portent for plans—such as those being drawn by governments in the U.S. and E.U.—to enhance local mining of lithium and other "green" minerals.

The worry is that the energy transition enhances well-worn patterns in which damaging mining is outsourced out to poorer nations where civil society is less equipped to challenge them. Two months after its Serbia mine was shut, Rio Tinto completed an $825 million purchase of a DLE-based lithium project in Salta, Argentina.

Many environmentalists think the fairest method to cut greenhouse gas emissions would be a more dramatic shift of consumption: we would need less lithium if we created fewer electric vehicles and instead depended more on public transport and traveled less in the first place.
That vision is unlikely to curb the mining growth ongoing in regions like Catamarca, however. That leaves communities uncomfortably dependent on businesses keeping to their new green promises—and the government holding them to account. If they don't, the battle against climate change, and the droughts and heatwaves it will bring, will be useless here, Guitián ads. "In the future, we'll have lithium, we'll have electric vehicles, but we won't have water,"

he argues. "We end up exactly at the same place."

CHAPTER FIVE

Argentina is currently at the forefront of a worldwide race for lithium. The nation, which alone accounts for 21 percent of the world's deposits, has only two mines in operation today, but 13 more are planned and dozens more are under consideration—the world's biggest lithium project pipeline. Argentina's embryonic boom represents a dramatic transition in the lithium market: a few years ago, lithium was a pretty specialist commodity, used to create glass, ceramics, and lubricants. Most of it was derived from well-established mines in Chile, Australia, and China. But with the global energy shift predicted to spark a 40-fold rise in lithium demand between now and 2040, according to the International Energy Agency, mining corporations are fighting to secure supply in less developed countries, such as Argentina and Bolivia, as well as Serbia and Mali. Many worried local populations in these places, notably the Argentina Puna, home to some 50,000 people, are destined to face the same destiny as those in Chile: having their resources pillaged and lands devastated to service the markets of wealthier countries—neocolonialism dressed up as a green revolution.

says Florentín Vasquez, a pleasant 38-year-old with a floppy black sun hat, standing with his two brothers beside a half-built adobe home in March 2022. "He claims that they're not going to need as much water as previous projects—that they have a novel approach that has never been tested before in Argentina."

A few kilometers away from the brothers' property, Lake Resources, the mining firm Promnitz started six years ago, is currently laying the framework for a salt lake lithium mine—dubbed Kachi. Using technology from California-based start-up Lilac Solutions, the business plans to start manufacturing lithium carbonate (the metal compound that battery makers purchase) in

2024, delivering 50,000 metric tons a year by 2025. Traditional lithium mines rely on a simple two-year-long evaporation process to separate lithium from the salty brine, allowing massive amounts of water to escape; by contrast, in a few hours, Lilac's technology can recover up to twice as much lithium and return "virtually all" of the salt water to its aquifer, according to Promnitz.

This sort of method is known as direct lithium extraction or DLE, and Kachi is one of the world's most sophisticated projects to employ it. The technique is untested, though. Even with major lithium producers researching DLE, including the world's biggest producer, North Carolina's Albemarle Corp., experts say it has failed to get from the lab to the field. Some investment organizations have raised severe reservations that Lake Resources will make it work at scale on its ambitious timeframe. The corporation has also endured internal upheaval: in June, Promnitz unexpectedly announced his departure. He tells TIME changes to the company's leadership were "anticipated" ahead of commencing construction at Kachi, while Lake's chair Stu Crow claims the resignation "was for strictly personal reasons." Despite the turbulence, Kachi remains a significant test case for DLE

This efficient technique is vital to scaling up lithium production and averting devastating bottlenecks in the energy revolution, hampering the battle against climate change. U.S. Energy Secretary Jennifer Granholm has termed it "a game changer" for the battery sector. David Snydacker, Lilac's CEO, argues it must succeed: "Conventional players have not been capable of supplying fresh supplies and the numbers necessary for electric vehicles," he argues. "So by 2030, either there's a disaster in the electric car market, or the lithium sector has been fundamentally transformed."

Working conditions at cobalt mines have inspired human-rights campaigners to call the ore "blood cobalt." In Chile, NGOs term places harmed by copper and lithium mining "sacrifice zones." If Kachi succeeds, it might help Argentina's lithium industry avoid a similar moniker. "If we're going to undertake an energy transition, we can't simply repeat the faults of the past," Promnitz adds. "We've got to do it better."

Bolivia's Marxist government, which possesses the world's biggest lithium resources but has dreaded the social and environmental effects of using them, seems to view DLE as the answer. In June, following a trial program investigating lithium recovery rates and water consumption, authorities declared that six DLE businesses, including Lilac, would be permitted to compete for lithium contracts there.

Argentina is currently at the forefront of a worldwide race for lithium. The nation, which alone accounts for 21 percent of the world's deposits, has only two mines in operation today, but 13 more are planned and dozens more are under consideration—the world's biggest lithium project pipeline. Argentina's embryonic boom represents a dramatic transition in the lithium market: a few years ago, lithium was a pretty specialist commodity, used to create glass, ceramics, and lubricants.

Most of it was derived from well-established mines in Chile, Australia, and China. But with the global energy shift predicted to spark a 40-fold rise in lithium demand between now and 2040, according to the International Energy Agency, mining corporations are fighting to secure supply in less developed countries, such as Argentina and Bolivia, as well as Serbia and Mali. Many worried local populations in these places, notably the Argentina Puna, home to some 50,000 people, are destined to face the same destiny as those in Chile: having their resources pillaged and lands devastated to service the markets of wealthier countries—neocolonialism dressed up as a green revolution.

But Stephen Promnitz, an Australian mining executive, informed the Vasquezes that he had the means to collect lithium while conserving their birthplace. "He was quite polite," says Florentín Vasquez, a pleasant 38-year-old with a floppy black sun hat, standing with his two brothers near a half-built adobe home in March 2022. "He claims that they're not going to need as much water as previous projects—that they have a novel approach that has never been tested before in Argentina."

A few kilometers away from the brothers' property, Lake Resources, the mining firm Promnitz started six years ago, is currently laying the framework for a salt lake lithium mine—dubbed Kachi. Using technology from California-based start-up Lilac Solutions, the business plans to start manufacturing lithium carbonate (the metal compound that battery makers purchase) in 2024, delivering 50,000 metric tons a year by 2025. Traditional lithium mines rely on a simple two-year-long evaporation process to separate lithium from the salty brine, allowing massive amounts of water to escape; by contrast, in a few hours, Lilac's technology can recover up to twice as much lithium and return "virtually all" of the salt water to its aquifer, according to Promnitz.

The economic incentives to roll out the technology couldn't be clearer. At the time, you couldn't power electric vehicles or store renewable energy without lithium. A frenzy to secure supply has sparked a roughly 500 percent surge in the price of lithium carbonate in the past 12 months—though experts think the pricing constraint may ease owing to a flood of new lithium investments.

Proponents argue DALE's speedier, more efficient technique is important to ramp up lithium production and avert fatal bottlenecks in the energy shift, hampering the battle against climate change. U.S. Energy Secretary Jennifer Granholm has termed it "a game changer" for the battery sector. David Snydacker, Lilac's CEO, argues it must succeed: "Conventional players have not been capable of supplying fresh supplies and the numbers necessary for electric vehicles," he argues. "So by 2030, either there's a disaster in the electric car market, or the lithium sector has been fundamentally transformed."

When construction is complete in two years, there will be additional extraction wells and covered tanks for the ion-exchange process vital to DLE's efficiency. "We'll put the brine in those tanks for only three hours," Promnitz adds, blinking at the sun behind his spectacles. Inside, lithium atoms will separate from the water molecules, and link instead with microscopic ion-exchange beads made by Lilac Solutions. The beads are then pulled out from the brine and cleaned with a strong acid to separate lithium chloride. Meanwhile the brine—around 800 metric tons of each metric ton of lithium carbonate produced—can be returned to the aquifer, Promnitz explains. In principle, the approach should avoid groundwater depletion as observed by communities in Chile. Some freshwater, however, is subsequently utilized in the process of converting lithium chloride into lithium carbonate, to be delivered to firms that build battery cathodes.

Energy Agency, mining corporations are racing to obtain supply in less developed regions, such as Argentina and Bolivia, as well as Serbia and Mali. Many worried local populations in these places, notably the Argentina Puna, home to some 50,000 people, are destined to face the same destiny as those in Chile: having their resources pillaged and lands devastated to service the markets of wealthier countries—neocolonialism dressed up as a green revolution.

But Stephen Promnitz, an Australian mining executive, informed the Vasquezes that he had the means to collect lithium while conserving their birthplace. "He was quite polite," says Florentín Vasquez, a pleasant 38-year-old with a floppy black sun hat, standing with his two brothers near a half-built adobe home in March 2022. "He claims that they're not going to need as much water as previous projects—that they have a novel approach that has never been tested before in Argentina."

A few kilometers away from the brothers' property, Lake Resources, the mining firm Promnitz started six years ago, is currently laying the framework for a salt lake lithium mine—dubbed Kachi. Using technology from California-based start-up Lilac Solutions, the business plans to start manufacturing lithium carbonate (the metal compound that battery makers purchase) in 2024, delivering 50,000 metric tons a year by 2025. Traditional lithium mines rely on a simple two-year-long evaporation process to separate lithium from the salty brine, allowing massive amounts of water to escape; by contrast, in a few hours, Lilac's technology can recover up to twice as much lithium and return "virtually all" of the salt water to its aquifer, according to Promnitz.

Flamingos were discovered in the saline lagoon near where Lake Resources' Kachi project will be constructed. Mining growth in the Andes might impact a sensitive habitat home to flamingoes and other creatures. Sebastián López Brach for TIME
This sort of method is known as direct lithium extraction or DLE, and Kachi is one of the world's most sophisticated projects to employ it. The technique is untested, though. Even with major lithium producers researching DLE, including the world's biggest producer, North Carolina's

Albemarle Corp., experts say it has failed to get from the lab to the field. Some investment organizations have raised severe reservations that Lake Resources will make it work at scale on its ambitious timeframe. The corporation has also endured internal upheaval: in June, Promnitz unexpectedly announced his departure.

He tells TIME changes to the company's leadership were "anticipated" ahead of commencing construction at Kachi, while Lake's chair Stu Crow claims the resignation "was for strictly personal reasons." Despite the turbulence, Kachi remains a vital test case for DLE.
The economic incentives to roll out the technology couldn't be clearer. At the present, you can't power electric cars or store renewable energy without lithium. A frenzy to secure supply has sparked a roughly 500 percent surge in the price of lithium carbonate in the past 12 months—though experts think the pricing

constraint may ease owing to a flood of new lithium investments. Proponents argue DALE's speedier, more efficient technique is important to ramp up lithium production and avert fatal bottlenecks in the energy shift, hampering the battle against climate change. U.S. Energy Secretary Jennifer Granholm has termed it "a game changer" for the battery sector. David Snydacker, Lilac's CEO, argues it must succeed: "Conventional players have not been capable of supplying fresh supplies and the levels necessary for electric vehicles," he adds. "So by 2030, either there's a disaster in the electric car market, or the lithium sector has been fundamentally transformed."

The Vasquez brothers pose near a half-finished farm home they are constructing. Sebastián López Brach for TIME
There's also a lot on the line for environmental justice. Climate supporters

have long been worried by the fact that extracting so-called green minerals required for decarbonization—lithium, cobalt, copper, and more—requires mining procedures that frequently damage ecosystems and injure people. In the Democratic Republic of the Congo, terrible working conditions in cobalt mines have inspired human-rights campaigners to call the resource "blood cobalt." In Chile, NGOs term places harmed by copper and lithium mining "sacrifice zones." If Kachi succeeds, it might help Argentina's lithium industry avoid a similar moniker. "If we're going to undertake an energy transition, we can't simply repeat the faults of the past," Promnitz adds. "We've got to do it better."

Bolivia's Marxist government, which possesses the world's biggest lithium resources but has dreaded the social and environmental effects of using them, seems to view DLE as the answer. In June, following a trial program investigating lithium recovery rates and water consumption, authorities declared that six DLE businesses, including Lilac, would be permitted to compete for lithium contracts there.

But amid the rush for lithium, Argentine environmentalists argue that a fairer future is far from assured. Most mining businesses aren't waiting for the adoption of cleaner tech. And projects like Kachi, which promise "cleaner lithium," have yet to show they can run without sapping freshwater supplies or affecting a little-understood environment. The uncertainty hangs on the Vasquez brothers, who have lived on this property for three generations, and expect to pass it on to their children one day. "People from outside may come and tell you, 'Don't worry nothing will happen,'" Florentín explains. "But we're the ones at risk."

To find a lithium mine in South America, look for the evaporation ponds: vast, vibrant blue-green pools of brine. Such rectangular ponds stretch dozens of square miles in Chile's Atacama desert, spilling tens of millions of metric tons of water into the air each year—at least 383.5 metric tons every metric ton of lithium carbonate produced, according to estimations by Argentine researchers. In Catamarca, the ponds are now an uncommon sight—for now: Argentina, which has suffered for decades with investor-spooking economic turmoil, has been hesitant to establish its lithium sector. But at least 14 projects are presently under investigation or building in this province alone.

Muerto salt lake, where U.S.-headquartered lithium miner Livent has been active for 20 years. Sebastián López Brach for TIME The Hombre\sKachi appears distinct from these typical mines. Early on a sunny morning in March, the contrasting hues of the location practically hurt your eyes: a white salt lake, like half-melted snowfall, stands at the foot of a black volcano—all framed by pink and orange mountains and the blue sky. One big red drill bores holes into the lake's salty crust to gather samples of brine from below. When construction is complete in two years, there will be additional extraction wells and covered tanks for the ion-exchange process vital to DLE's efficiency. "We'll put the brine in those tanks for only three hours," Promnitz adds, blinking at the sun behind his spectacles. Inside, lithium atoms will separate from the water molecules, and link instead with microscopic ion-exchange beads made by Lilac Solutions. The beads are then pulled out from the brine and cleaned with a strong acid to separate lithium chloride. Meanwhile the brine—around 800 metric tons of each metric ton of lithium carbonate produced—can be returned to the aquifer, Promnitz explains. In principle, the approach should avoid groundwater depletion as observed by communities in Chile. Some freshwater, however, is subsequently utilized in the process of converting lithium chloride into lithium carbonate, to be delivered to firms that build battery cathodes.

Ponds storing fresh water at the Kachi location. Sebastián López Brach for TIME

Kachi is the most advanced project for Lilac Solutions, whose backers include BMW, as well as Breakthrough Energy Ventures, an investment group supported by Bill Gates and Jeff Bezos. Lilac says it has completed a pilot project using the same beads and ion-exchange process somewhere "in the western U.S." that produced 25 metric tons of lithium carbonate equivalent a year—that's enough to make batteries for roughly 400 Teslas, according to estimates produced in 2015 by analysts at Goldman Sachs. A somewhat bigger pilot will be built this year at Kachi, before commercial-scale production starts in 2024, providing Lilac an opportunity to establish itself in the worldwide market. In a vote of confidence in April, the Ford Motor Co. inked a nonbinding deal with Lake Resources to acquire 25,000 metric tons of lithium carbonate yearly from Kachi. Japanese business Hanwa has signed a similar MoU for the other half.

It is likely simply a fortunate accident that more efficient—and consequently profitable—lithium-extraction processes also have a lower land and water impact than previous ones. Nevertheless, DLE has attracted green-minded investors, including the U.K. government's foreign credit agency, which is in discussions to contribute 70 percent of the finance for Kachi's development. Snydacker said Lilac's more sustainable technology "can allow project developers to avoid the backlash" that some are currently witnessing from local communities to lithium projects. Chilean mining corporations have faced expensive legal disputes over their usage of brine. In Argentina, anti-mining rallies are prevalent in communities in Catamarca and nearby Jujuy and Salta provinces, making national headlines about the risks to local water sources.

Argentine environmentalists are interested, but not reassured, by Lake and Lilac's promises to cut water usage. "They are guaranteeing us it would have a reduced impact," says Patricia Marconi, a Catamarca-based researcher with the YU CHAN Foundation, a regional conservation organization. "But they don't have anything published." She alleges Lake has refused to share information with her and her coworkers. The corporation has not yet produced its environmental impact assessment study, generating doubt. (Lake Resources chair Stu Crowe tells TIME the study is still being completed and is slated to be released in the third quarter of 2022.)

CHAPTER SIX

Brines recovered from the salt lake in Kachi are kept in a well.

Eva Vasquez was born in the region where the Kachi project is being created. Sebastián López Brach for TIME

Marconi's worries revolve around two issues: First, how will reinjecting vast volumes of brine into the aquifer alter the geological formations below the surface? In a 2018 publication, a group of Argentina-based academics stated that reinjecting the previously lithium-rich brines into the lakes is "a very risky oversimplification" of the possible environmental consequences of the practice. According to the report, treating the brines to remove lithium might change their pH and introduce trace quantities of foreign chemicals. And firms may have to inject leftover brine at other sites than where they extracted it, simply to prevent diluting the lithium concentration where they're extracting.

The second issue is, how much fresh water will be used in the latter phases of the process? Marconi cautions that this water will likely be so-called fossil water—drawn from aquifers locked below for thousands of years that are not supplied quickly enough by today's rains to be restored.

All of this, Marconi argues, will have an unknown influence on the little-studied geology and sensitive biology of the salt lakes. In their zeal for lithium, she argues Argentina's national and provincial governments are neglecting to complete the research necessary to predict the implications of huge mining development, turning the country into a "free for all" for the lithium business. "If we were truly taking seriously the principle of not engaging in environmental systems that we don't understand, there wouldn't be 20 corporations investigating in Catamarca. There would be 20 study teams examining what's going to happen," she adds. "Because the damages are irreversible."

Catamarca has caused us to question DALE's environmental claim. Three hours north of Kachi—a teeth-chattering journey over rough mountain roads—the yellows and greens of the Puna's summer flora are abruptly broken by a startling expanse of black ground. This is the valley of the Trapiche River, a water supply for the huge Hombre Muerto salt lake. In 1997, Livent, a Philadelphia-based lithium mining company—a critical supplier for both Tesla and BMW—built a small dam at the point where the river flows into the salt lake. The dam concentrates the freshwater for use in Livent's mine, which now can generate up to 20,000 metric tons of lithium per year. On TIME's visit in March—the last month of the Puna's wet season—a trickle of water a few feet wide ran past the dam, through a parched, blackened meadow.

The Trapiche River's course is blocked by a dam erected by Livent near the Hombre Muerto salt lake. Sebastián López Brach for TIME

The operation is Argentina's oldest lithium mine—and it's also the only one in the western hemisphere to deploy a variant of DLE at scale. Its procedure is a hybrid: lithium brines are allowed to evaporate in pools but for "significantly less time," according to Livent than in standard methods, decreasing saltwater loss. The brine is then processed through a DLE process and subsequently, Livent adds, "most" of the salt water is returned to "the neighboring Salar habitat." Later, fresh water from the Trapiche River is used to separate the lithium. The firm did not release any data on brine consumption, but asserts it has "not contributed to a drop in brine or water [in the two decades it has worked] at the Salar."

Román Guitián blames Livent for the valley's ruin. Guitián grew up adjacent to the river, in a tiny Indigenous hamlet made up of his family and a few others. Before the mining began when he was 17, they used to get salt from Hombre Muerto, and farmed llamas, goats, and sheep on the valley's vegetation, says Guitián, standing at the edge of the salt lake next to a beat-up 4×4 that he employs to take visitors around the highlands. "It was lovely. But now there are no animals since it's completely dry."

Signs strewn along the river proclaim an initiative to rehabilitate the valley via reforestation and new irrigation systems, which Livent initiated last year with a regional NGO, the Eco-Conciencia Foundation. And yet, in early March, during a ceremony at Hombre Muerto attended by Argentine politicians and Tesla executives,

Livent announced intentions to increase the plant's lithium production capacity by the end of 2023. The business is also planning two more extensions, aiming to raise its total capacity to 100,000 metric tons by 2030. In investment documents, Livent promises that the company would combine "reuse" and "recycling" to decrease its fresh water usage in the future. But it also says that the later stage of expansion will involve "a more conventional pond evaporation-based process."

During TIME's visit, workers were busy excavating a pipeline from the Hombre Muerto facility to another river, Los Patos, some 10 miles distant. "They ruined one river and now they're going to damage another," adds Guitián. He is actively seeking formal recognition of an Indigenous community, Atacameños del Altiplano, created with a few dozen other local people. The designation would provide the community a constitutional right to participate in the management of natural resources in their area. He'd use it to safeguard the environment from "irresponsible mining" methods, he claims. "If the day comes when we don't have water, we'll have to emigrate.".

Catamarca governor Raúl Jalil believes the province has learned from the repercussions on the Trapiche River. "There are things which maybe went the wrong way in the past, but we are fixing those," he adds. "We are taking greater control now." Companies are now forced to undertake monthly environmental monitoring, and if concerns crop up, projects will be stopped, Jalil said. The Levant claims it has already constructed monitoring stations at both the Trapiche and Los Patos rivers "to check water levels, recharge rate, and water chemistry to enable us to utilize water sustainably."

Jalil, though, says he does not seek to limit the number of lithium projects permitted in the province, or block conventional, water-intensive technologies in future mines, unlike Bolivia's government. "All enterprises, from agriculture to tourism, have an environmental effect, and we can't overhaul the entire global energy system without mining," he argues. "The path ahead is to lessen the effect more and more, via technology, innovation—at the same time as [continued] extraction." He wants Kachi to be "a leading case" there.

Promnitz thinks Kachi will utilize water more effectively than Livent has: Livent's worldwide freshwater consumption for 2020—based mostly on the Hombre Muerto project, its sole operating lithium carbonate mine—was 72.9 metric tons for every metric ton of lithium carbonate equivalent (LCE) produced. In the lab, Lilac claims it consumed 18 metric tons of water for every metric ton of LCE. Lake believes the rate at Kachi will be much below that, and that more water savings would come from plans to utilize brackish and recycled water.

Promnitz, who started his career as a geologist for mining industries, also thinks the brine reinjection procedure is not as hazardous as environmentalists worry. The process is similar to one that has been used for decades to enhance oil recovery in the U.S. shale sector, he says, except that the looser sediments of the salt lakes, compared to the tighter formations under oil fields in Texas and Louisiana, should make it less important where the brine is reinjected. "It's not like [the brine] lives in just one specific structure. We're merely pulling the lithium out, and putting back precisely the same brine that was there before."

In discussion, Jalil and other local leaders in Catamarca seem considerably more concerned with mining's economic possibilities than its environmental hazards. Argentina's northwest has long been a backwater, getting little investment from either Buenos Aires or elsewhere. Politicians view mining as an opportunity to alter that: corporations such as Livent have paid for roads and bridges in areas that were until recently inaccessible during poor weather; Lake is in negotiations to recruit local people and contract services, such as laundries, with local businesses. Though leery of unmet promises, the mayors of both El Peñón and Antofagasta de la Sierra, Kachi's two nearest towns, express variants of the same phrase: "If the firm expands, the municipality should too."

It's reasonable that local authorities would adopt that approach, says Juan Carrizo, head of the Eco-Conciencia Foundation, which strives to "resolve socio-environmental conflicts" surrounding mining. "It's simple to protect the environment from areas like Buenos Aires, where you have the internet, roads, gas, and transport," he argues. "But here, the growth of the community is also on the table."

A discussion over mining plays out on the streets of Antofagasta de la Sierra. Proudly displayed bumper stickers say "I'm a huge fan of mining," while graffiti begs "hands off our land and water." Florentín, whose property stands near the Kachi site, feels conflicted. He argues that if environmental concerns did develop from the project, he wouldn't be sure what to do. "We can attempt to push the corporation to leave … but I don't believe we want to do that," he replies softly. "There are a lot of individuals around here that need a job, and they come and tell me that. So I feel sort of cornered."

We are experiencing a worldwide mineral mining boom. Carbon-reducing technology like wind turbines, solar panels, and electric automobiles demand a bigger volume and a more diversified variety of minerals than their dirtier equivalents. Already, since 2010, the International Energy Agency stated, "the average quantity of minerals required for a new unit of power production capacity has climbed by 50 percent as the percentage of renewables has risen." That provides the global mining sector a shot at a rebrand: from environmental villain to climate savior.

But in an age of heightened environmental consciousness, people are also fighting back against mining operations. In January, following weeks of huge demonstrations, the government in Serbia—thought to be home to Europe's greatest lithium deposits—shelved a $2.4 billion project sponsored by mining giant Rio Tinto, which could have given 90 percent of Europe's current lithium demands. Though Rio Tinto is now seeking to continue discussions with the government, which won re-election in April, the experience is a terrible portent for plans—such as those being drawn by governments in the U.S. and E.U.—to enhance local mining of lithium and other "green" minerals.

The worry is that the energy transition enhances well-worn patterns in which damaging mining is outsourced out to poorer nations where civil society is less equipped to challenge them. Two months after its Serbia mine was shut, Rio Tinto completed an $825 million purchase of a DLE-based lithium project in Salta, Argentina.

Many environmentalists think the fairest method to cut greenhouse gas emissions would be a more dramatic shift of consumption: we would need less lithium if we created fewer electric vehicles and instead depended more on public transport and traveled less in the first place.

That vision is unlikely to curb the mining growth ongoing in regions like Catamarca, however. That leaves communities uncomfortably dependent on businesses keeping to their new green promises—and the government holding them to account. If

they don't, the battle against climate change, and the droughts and heatwaves it will bring, will be useless here, Guitián ads. "In the future, we'll have lithium, we'll have electric vehicles, but we won't have water," he argues. "We end up exactly at the same place."
'Lithium Fields' on the Salar de Atacama salt flats in northern Chile.
Lithium mining sites in South America have been recorded by an airborne photographer in spectacular high definition.
But although the photographs may be magnificent to look at, they depict the dark side of our quickly changing world.

Lithium provides a path out of our dependency on fossil fuel production. As the lightest known metal in the world, it is now extensively employed in electric equipment from mobile phones and computers to vehicles and airplanes.

Lithium-ion batteries are most renowned for powering electric cars, which are anticipated to account for up to 60 percent of new automobile sales by 2030. The battery of a Tesla Model S, for example, requires roughly 12 kg of lithium.

These batteries are the key to lightweight, rechargeable power. As it stands, demand for lithium is enormous and many argue it is needed to switch to renewables.
However, this doesn't come without a cost - mining the chemical element may be hazardous to the ecosystem.

German aerial photographer Tom Hegen specialized in recording the footprints humans leave on the earth's surface. His work presents an overview of regions where humans harvest, refine and use materials with his current series revealing the "Lithium Triangle."

Lithium provides a path out of our dependency on fossil fuels - it is most renowned for powering electric cars.
This area rich in natural riches may be located where the borders of Chile, Argentina, and Bolivia meet. And nearly a fourth is kept on the Salar de Atacama salt plains in northern Chile.

"To get the vast mining activities in the frame, I leased tiny airplanes and flew high above them," Hegen adds.
His photos of the Soquimich lithium mine in the Atacama desert, managed by prominent mining company Sociedad Química y Minera (SQM), are part of his new project, The Lithium Series I.

Why are the fields so colorful?
The brilliant colors of the lithium fields, or ponds, are created by differing quantities of lithium carbonate. Their color may vary from a pinkish white to turquoise, to a very concentrated, canary yellow.

But when we think of extraction, we think of fossil fuels like coal and gas. Unfortunately, lithium also falls under the same umbrella, despite paving the way for an electric future. Lithium may be regarded as the non-renewable material that makes renewable energy feasible - frequently hailed as the next oil.

Lithium mining always affects the soil and creates air pollution.
According to research by Friends of the Earth (FoE), lithium mining always affects the soil and causes air pollution. As demand climbs, the mining consequences are "increasingly harming communities where this hazardous extraction is happening, jeopardizing their access to water," says the research.

The salt flats in South America where lithium is discovered are situated in dry countries. In these locations, access to water is vital for the local populations and their livelihoods, as well as the native flora and wildlife.

In Chile's Atacama salt flats, mining consumes, contaminates, and diverts precious water supplies away from residents.

CONCLUSIONS

Approximately 2.2 million gallons of water are required to manufacture one ton of lithium.

The manufacturing of lithium via evaporation ponds consumes a lot of water - roughly 21 million liters per day. Approximately 2.2 million gallons of water are required to manufacture one ton of lithium.

"The mining of lithium has produced water-related problems with various groups, such as the village of Toconao in the north of Chile," the FoE study states.

Tom Hegen
Hegen's photographic project, The Lithium Series I.
Tom Hagen
Scientists claim chemical pollution has now breached the permissible limit for mankind

Blue rivers with water as strong as bleach: The 'destructive' influence of fast fashion in Africa

This is the greatest nation in Europe for driving an electric vehicle

Where are additional lithium hotspots across the world?

The increased interest in lithium has seen the world's largest-known deposits climb dramatically. There are roughly 80 million tonnes of identified deposits worldwide as of 2019, according to the US Geological Survey (USGS).

After South America (chiefly Bolivia, Chile, and Argentina) the next greatest lithium-producing nation is the United States, followed closely by Australia and China.

In 2019, lithium exports from Australia are believed to have amounted to over $1.6 billion (€1.3bn).

Much like previous struggles and conflicts over gold and oil, governments are battling for control over minerals like lithium - since this might help them gain economic and technical superiority for decades to come.

Other nations with lower reserves include Zimbabwe, Brazil, and, the lone European country, Portugal.

Lithium mining has proven especially contentious lately in Portugal, with the town of Pinhel now ready to pursue an injunction to prohibit the exploration. Portuguese locals have continually agitated against the rare metal's extraction, alleging enormous environmental repercussions. But the government has given the green light to the extraction of the "white gold" in six different regions.

95 percent of the local people have opposed these proposals, despite the mining company's assertions that the ore's

exploitation would provide roughly 800 employment for locals.

So should we stop harvesting lithium for batteries?
A similar assessment produced in 2021 by the charity Bebe (Bienaventurados de Pobres) similarly lists water as a significant risk for lithium mining activities.

It states that not enough study has been done on the possible pollution of water and "activity must be halted until studies are available to accurately evaluate the severity of the damage."

Gleb Yushin, a professor at the School of Materials and Engineering at Georgia Institute of Technology, US, says that new battery technology has to be created utilizing more common, environmentally-friendly materials. His study is published in the journal Nature, with co-authors including Kostiantyn Turcheniuk.

As supplies of lithium and cobalt will not fulfill future demand, proposed elements to concentrate on instead include iron and silicon.

Researchers like Yushin are working on novel battery options that would replace lithium and cobalt (another dangerous metal) with less toxic and more readily available materials. As supplies of lithium and cobalt will not fulfill future demand, proposed elements to concentrate on instead include iron and silicon.

Unlike lithium-ion batteries, iron flow batteries are also cheaper to make, renewable energy veteran Rich Hossfeld told Bloomberg recently, in a story headed 'Iron battery innovation might eat lithium's lunch.

"We urge materials scientists, engineers, and funding organizations to prioritize the research and development of electrodes based on plentiful elements," asserts Yushin.

"Otherwise, the roll-out of electric automobiles would halt within a decade."

www.ingramcontent.com/pod-product-compliance
Lightning Source LLC
Chambersburg PA
CBHW070300220526
45465CB00004B/1676